I M A G E S

Materials

Karen Bryant-Mole

Heinemann

First published in Great Britain by Heinemann Library, an imprint of
Heinemann Publishers (Oxford) Ltd, Halley Court, Jordan Hill, Oxford OX2 8EJ

MADRID ATHENS PARIS FLORENCE PRAGUE WARSAW
PORTSMOUTH NH CHICAGO SAO PAULO SINGAPORE TOKYO
MELBOURNE AUCKLAND IBADAN GABORONE JOHANNESBURG

Designed by Jean Wheeler
Commissioned photography by Zul Mukhida
Printed in Hong Kong

00 99 98
10 9 8 7 6 5 4 3 2

ISBN 0 431 06290 0

British Library Cataloguing in Publication Data
Bryant-Mole, Karen
 Materials. – (Images Series)
 I. Title II. Series
 531.11

**Some of the more difficult words in this book are
explained in the glossary.**

Acknowledgements
The Publishers would like to thank the following for permission to reproduce photographs. Chapel Studios;
5 (right), 10 (right), 17 (left), Positive Images; 10 (left), 11 (left), Tony Stone Images; 4 (left) Larry Ulrich,
5 (left) Bob Thomas, 11 (right) Roslav Szaybo, 16 (left) Nicholas DeVore, 16 (right) Jean-Marc Truchet,
17 (right) Peter Correz, Zefa; 4 (right).

Every effort has been made to contact copyright holders of any material reproduced in this book. Any omissions
will be rectified in subsequent printings if notice is given to the Publisher.

Contents

Liquids

Liquids are usually wet.

Water is a liquid.

Paint and
bubble mixture
are liquids, too.

Solids

Solids have a shape that usually stays the same.

Can you think of
some more solids?

Plastic

Plastic is waterproof.

It is strong
and light.

Plastic can be
made into all
sorts of shapes!

Wood

Wood comes from trees.

It is often made
into furniture
and buildings.

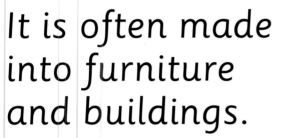

Wool

Wool comes from sheep.

It has to be spun
and then dyed.

It can be
used to make things
like this woolly hat.

Metal

There are many different types of metal.

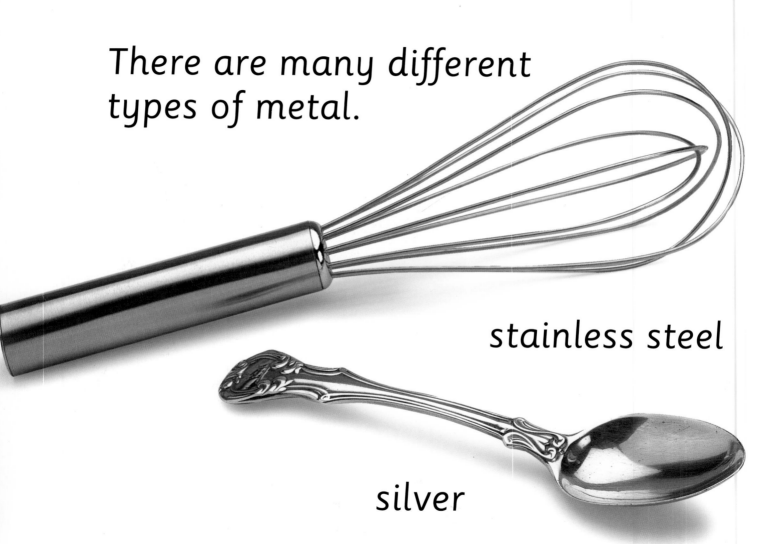

stainless steel

silver

brass

gold

Fabric

Some fabrics, like cotton and silk, are made from natural materials.

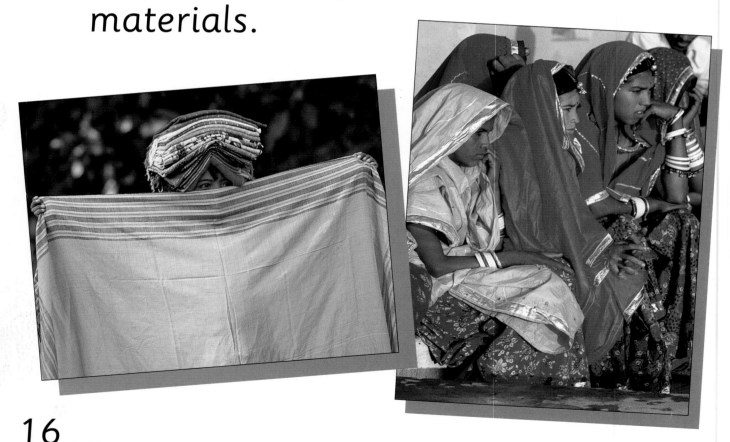

Other fabrics, like polyester and nylon, are synthetic.

Glass

Glass can be coloured or clear.

You can usually see
through clear glass.

Paper

some books

some tissues

a birthday card

a paper plate

All of these things
have been made
from paper.

21

Clay

Clay can be made into lots of shapes.

Glossary

dyed   made into a particular colour
material   what an object is made from
spun   pulled out and twisted
synthetic   not a natural material
waterproof   won't let water through

Index